CONTENTS

CHAPTER ONE

WEATHER AROUND US

Imagine you're outside on a sunny day. It's bright and warm, and the sky is clear. Suddenly, everything changes. Dark clouds roll in and make the sky very dark. The temperature drops several **degrees**, the wind blows stronger, and soon, heavy rains soak everything around you. This situation isn't that uncommon. Weather is unpredictable, which means it can change without warning.

Even in places where the weather is normally sunny and warm, bad weather can strike and cause people to seek shelter.

SPOTLIGHT ON WEATHER AND NATURAL DISASTERS

WEATHER AND CLIMATE AROUND THE WORLD

SARAH MACHAJEWSKI

PowerKiDS press™

NEW YORK

Published in 2019 by The Rosen Publishing Group, Inc.
29 East 21st Street, New York, NY 10010

Editor: Elizabeth Krajnik
Cover Design: Michael Flynn
Interior Layout: Tanya Dellaccio

Photo Credits: Cover (background) Cammie Czuchnicki/Shutterstock.com; cover nevodka/Shutterstock.com; p. 4 Drew McArthur/Shutterstock.com; p. 5 Nick Starichenko/Shutterstock.com; p. 7 Ellen Bronstayn/Shutterstock.com; p. 8 Peter Gudella/Shutterstock.com; p. 9 elnavegante/Shutterstock.com; p. 10 Bloomberg/Getty Images; p. 11 Kazakova Maryia/Shutterstock.com; p. 12 Courtesy of the Library of Congress; p. 13 Everett Historical/ Shutterstock.com; p. 15 GREG WOOD/AFP/Getty Images; p. 16 Jessica Moore/Cultura/Getty Images; p. 17 LAURA BUCKMAN/AFP/Getty Images; p. 18 https://commons.wikimedia.org/w/index.php?title=File:30_Jan_1977_Tonawanda,_NY_(Parker_Blvd_between_Brompton_Rd_and_Ellicott_Creek_Rd).jpg&oldid=190161498; p. 19 Ira Block/National Geographic/Getty Images; p. 21 Russell Curtis/Science Source/Getty Images.

Cataloging-in-Publication Data

Names: Machajewski, Sarah.
Title: Weather and climate around the world / Sarah Machajewski.
Description: New York : PowerKids Press, 2019. | Series: Spotlight on weather and natural disasters | Includes glossary and index.
Identifiers: LCCN ISBN 9781508169222 (pbk.) | ISBN 9781508169208 (library bound) | ISBN 9781508169239 (6 pack)
Subjects: LCSH: Weather--Juvenile literature. | Climatology--Juvenile literature.
Classification: LCC QC981.3 M33 2019 | DDC 551.5--dc23

Manufactured in the United States of America

CPSIA Compliance Information: Batch #CS18PK For further information contact Rosen Publishing, New York, New York at 1-800-237-9932.

From bright sun to heavy rain and hurricane winds to snowy blizzards, Earth experiences many kinds of weather. Under normal conditions, weather is a **phenomenon** that affects how we live, what we wear, and where we go. Extreme weather, on the other hand, is an **extraordinary** display of nature's power at work. These events happen around the world. Weather is around us at all times. Let's find out more about it!

CHAPTER TWO

ALL ABOUT THE ATMOSPHERE

What is weather, and what creates it? To begin with, weather is the conditions in an area over a short period of time. It's created by the activity of Earth's atmosphere, which is the whole mass of air surrounding Earth. The atmosphere is important because it traps the heat of the sun, keeping our planet warm. It plays a big role in the **water cycle**, and it's where weather conditions occur.

The atmosphere is made of a mix of gases—mostly nitrogen, oxygen, argon, and carbon dioxide. There's a lot of energy in the atmosphere. Other forces, including pressure, moisture, and heat, are at work in the atmosphere. When these forces work together to create weather conditions, we experience them as **precipitation**, heat, cold, **humidity**, dryness, and wind.

The atmosphere reaches from Earth's surface to the edge of space. There's less oxygen the higher you go in the atmosphere, which means life can't survive there.

LAYERS OF EARTH'S ATMOSPHERE

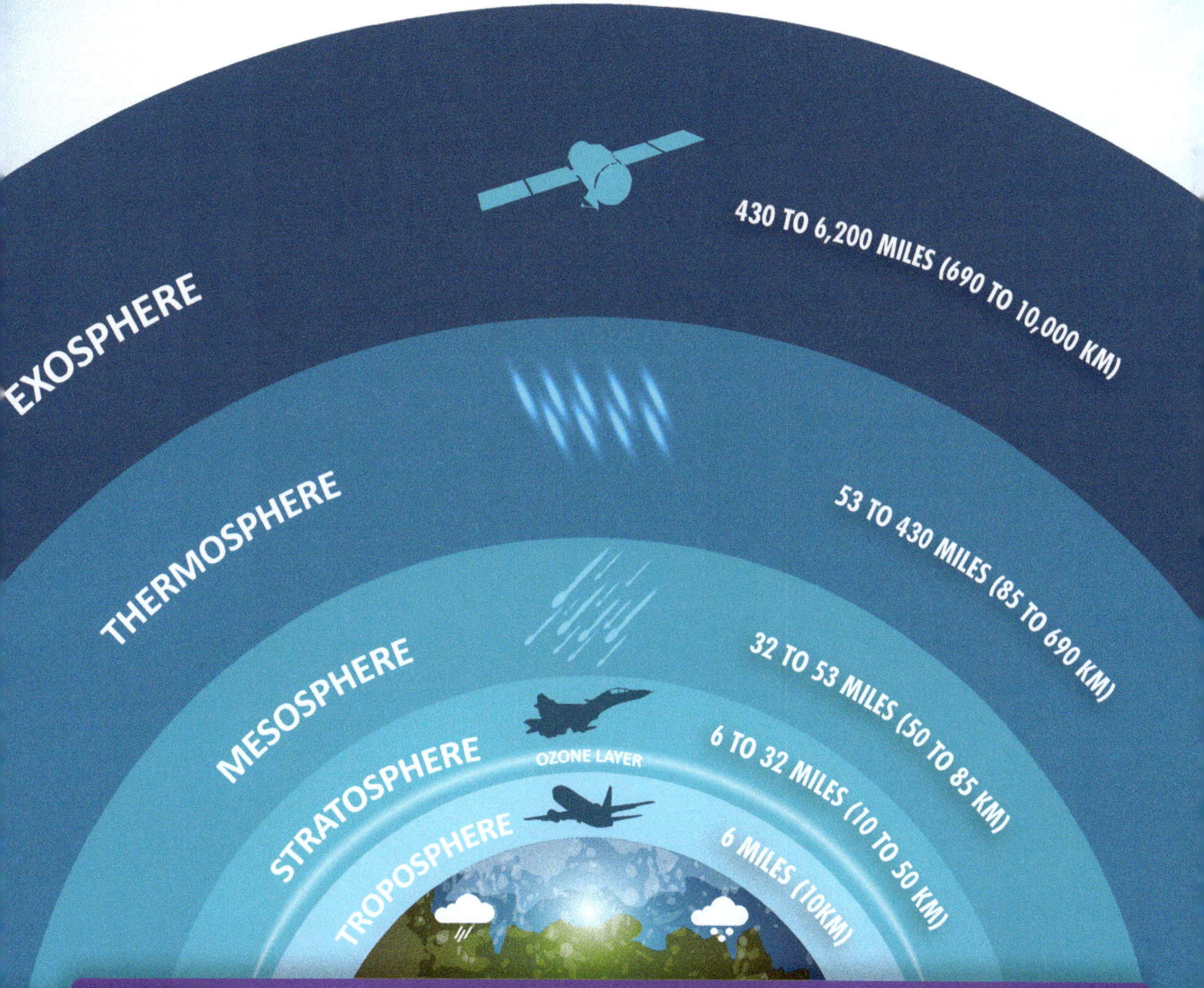

CHAPTER THREE

UNDERSTANDING CLIMATE

Climate and weather are different things. The climate of an area is the average weather conditions there over a long period of time. There are five main climate groups. Polar climates exist at Earth's poles and in **tundra** regions. Dry climates occur in areas with little water, such as deserts. **Tropical** climates exist close to the equator. Mild climates and continental climates have a mix of cool and warm weather.

Climate affects what can grow in an area, which is why types of plant and animal life vary around the world. Plants that grow in tropical climates can't survive in polar climates.

While weather can change by the minute or by the hour, climate changes very slowly—over tens, hundreds, and thousands of years. Earth's climates have always been in a state of change, but climate change has occurred more rapidly in the last 100 years due to human activity.

CHAPTER FOUR

THE WATER CYCLE AND PRECIPITATION

Precipitation is one of the most visible parts of weather. It's created as water moves through the water cycle. The water cycle plays a large role in Earth's weather patterns.

Evaporation is one step of the water cycle. In evaporation, the sun heats water on Earth's surface and this water changes from a liquid into a gas. This gas, called

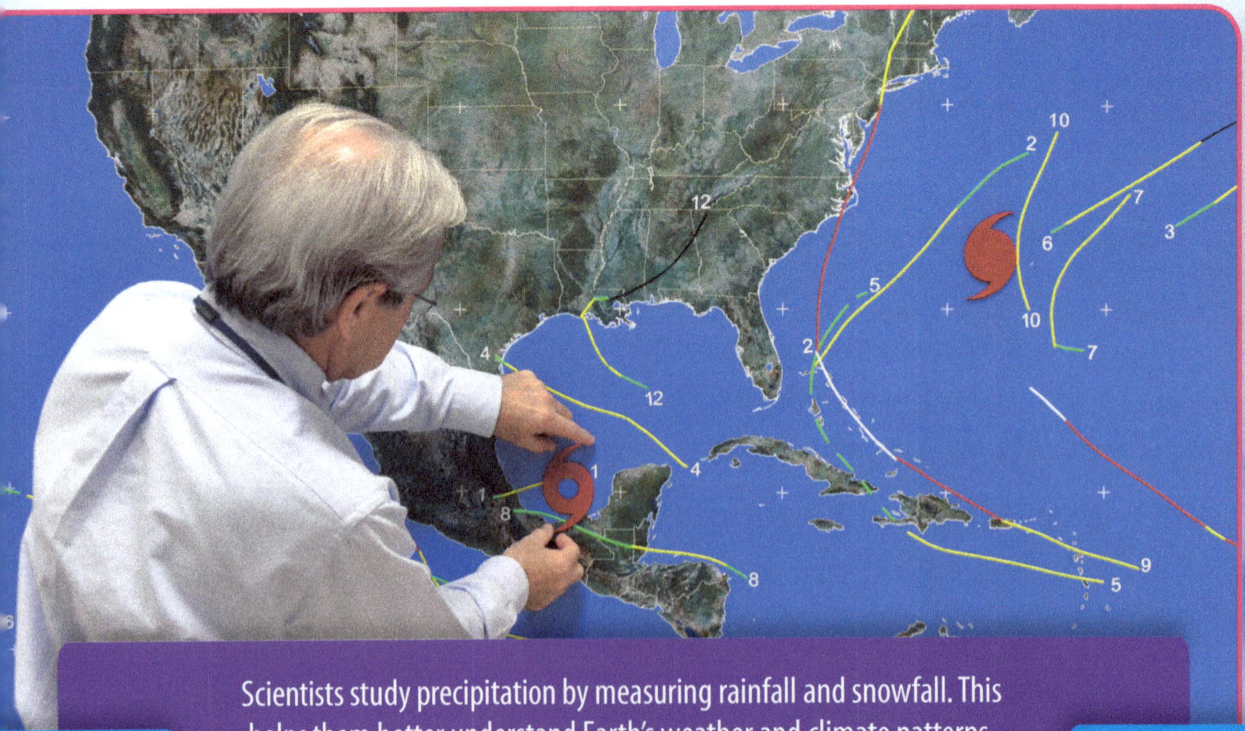

Scientists study precipitation by measuring rainfall and snowfall. This helps them better understand Earth's weather and climate patterns.

THE WATER CYCLE

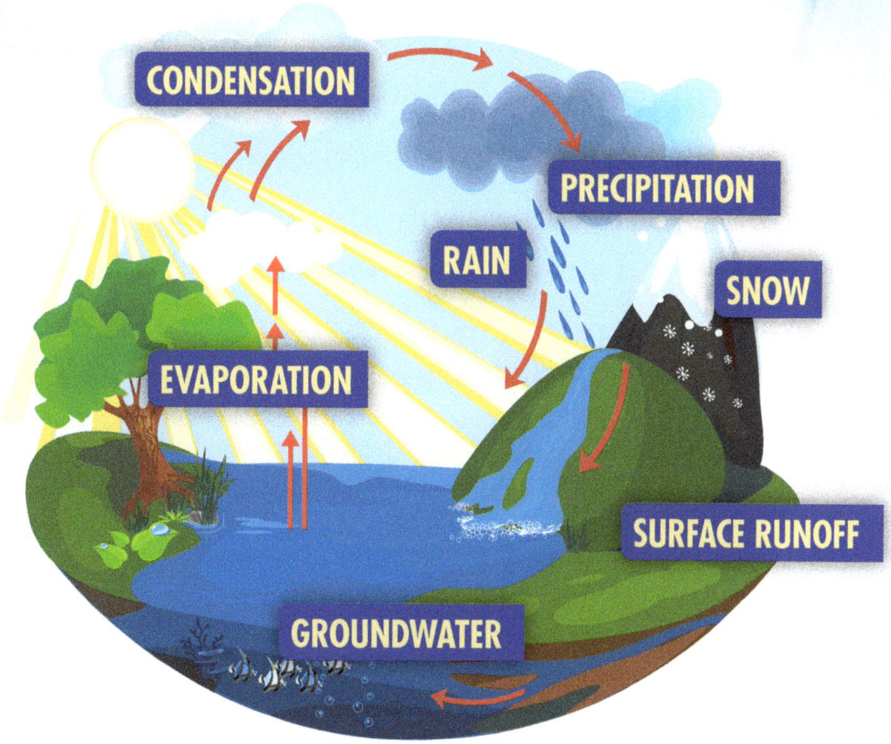

water vapor, rises into Earth's atmosphere. Water from trees and plants also becomes part of the atmosphere through a process called transpiration.

Condensation, or the process of water vapor cooling and turning into a liquid, is another step of the water cycle. Clouds are formed through condensation. When a cloud can't hold more water, the water falls to Earth as precipitation.

CHAPTER FIVE

FLASH FLOODS

A good soaking of rain is very important for life on Earth. But what happens when there's too much of it? Heavy rainfall can cause floods, which occur when more rain falls than the ground and waterways can hold. Rivers and streams overflow, and water collects in areas that are normally dry. In cities, floods can bring several feet of standing water to streets and roads.

The Johnstown Flood of 1889 was one of the deadliest floods in U.S. history. Heavy rainfall in the area caused a large dam to fail and flood the area.

Flash floods are periods of rapid flooding. They're the most dangerous type of flood because they happen quickly, without warning. Flash floods are the most common weather-related cause of death in the United States. Scientists are working hard to fix this. Using advanced **technology**, scientists make rainfall **estimates** to forecast these natural disasters and give people time to evacuate, or leave a place of danger. This use of science and technology helps save lives.

CHAPTER SIX

A LACK OF WATER

Heavy precipitation is one weather problem people must deal with, but what about when there's very little or no precipitation for a long period of time? This kind of weather is called a drought.

Droughts are caused by unusual weather patterns that affect how much precipitation an area receives and for how long. Short-term droughts can last for weeks or a few months. Sometimes, however, droughts can last for many months or years. This type of drought is called a long-term drought. It occurs when weather patterns become stuck over a certain area.

Climatologists, or scientists who study climate, work hard to keep track of temperatures, how much rain areas get, levels in bodies of water, moisture levels in soil, and how thick the snow is in any nearby mountains. All of this is necessary to understand droughts.

In times of long-term drought, **engineers** may decide the best and fastest way to get water to affected areas. This image was taken in 2005 at Pejar Dam near the city of Goulburn in New South Wales, Australia. This area had experienced four years of drought.

CHAPTER SEVEN

WILD WINDS

Wind is a part of weather we usually can't see, but we can definitely feel it. Wind is caused by air temperature and air pressure. Differences in temperature and pressure cause air **molecules** to move—fast or slow, up or down. As air moves from areas of high pressure to areas of low pressure, it bumps into objects, such as people, and we feel it as wind.

A tornado can destroy everything in its path. In 2015, a tornado hit Rowlett, Texas, and left behind large amounts of damage. It killed at least 11 people.

Wind can be harmless. However, it can also be harmful. Tornado winds can cause a lot of damage. A tornado is a narrow, **violently** spinning column of air that reaches from a thunderstorm to the ground. We can see a tornado if it has a funnel cloud, which is a cloud made of water droplets, dust, and other things the wind picks up. The worst tornado winds can reach up to 250 miles (402.3 km) per hour.

CHAPTER EIGHT

TROPICAL CYCLONES AND BLIZZARDS

Tropical cyclones—called hurricanes when they form over the Atlantic Ocean or eastern Pacific Ocean, typhoons over the western Pacific Ocean, and cyclones over the Indian Ocean—are large, powerful, and destructive storms with winds faster than 74 miles (119.1 km) per hour, rain, thunder, and lightning.

In 1977, a blizzard hit Buffalo, New York. While the region is used to dealing with the snow and cold, winds blew the snow into piles taller than houses—trapping many people inside for days.

Hurricanes form over warm ocean waters near the equator, in spots where warm air rises and creates an area of low pressure below. Air from the surrounding high-pressure areas fills the low-pressure space. The new body of air warms and rises, all the while spinning and growing larger.

A blizzard is when falling snow or blowing snow with winds of 35 miles (56.3 km) per hour or greater causes visibility of a quarter mile (0.4 km) or less for three hours or longer. This combination of snow and low visibility is very dangerous. When a blizzard hits an area, people should stay indoors.

CHAPTER NINE

ERUPTIONS, EXPLOSIONS, EARTHQUAKES

Volcanic eruptions are natural disasters that have changed landscapes and weather patterns around the world. A volcano is an opening in Earth's surface through which partly melted rock, called magma, can escape. Some volcanic eruptions are violent, shooting ash, rocks, and lava (magma above the surface) into the air as if an explosion has gone off. Other volcanic eruptions are calmer, with lava flowing from the opening.

Some eruptions are so powerful that they affect weather and climate. The ash, gas, and rocks thrown into the air can cause storms, affect winds, and block out the sun, causing temperatures to cool.

Earthquakes are caused by activity deep within Earth. When an earthquake occurs, the ground and everything on top of it shakes. Earthquakes can cause volcanic eruptions.

Seismologists, or scientists who study earthquakes, measure earthquakes with tools called seismographs, which detect and record vibrations, or shaking movements, from within the earth.

CHAPTER TEN

UNDERSTANDING CLIMATE CHANGE

No matter where you live or what time of year it is, weather is happening right now in your area. Weather is a part of our daily lives and affects how we live. But it's important to understand that how we live affects weather, too.

Scientists are growing more concerned about the effect human activity has on Earth's climate. **Evidence** shows that Earth's average temperature is rising, causing oceans to warm, glaciers and ice caps to melt, and sea levels to rise. All of this can change weather patterns. For example, scientists are observing more heavy rainfall than ever before.

Scientists are still studying our climate and the changes that affect us. In the meantime, it's important to treat Earth with respect so that it remains a healthy home for all the people, animals, and plants that live here.

GLOSSARY

degree (dih-GREE) A unit for measuring temperature.

engineer (en-juh-NEER) Someone who plans and builds machines using math and science.

estimate (EH-stuh-muht) A guess you make based on the information you have about something.

evidence (EH-vuh-duhnts) Something that shows that something else exists or is true.

extraordinary (ik-STROR-duh-nehr-ee) Very different from what is normal or ordinary.

humidity (hyoo-MIH-duh-tee) The amount of moisture in the air.

molecule (MAH-lih-kyool) The smallest possible amount of something that has all the characteristics of that thing.

phenomenon (fuh-NAH-muh-nahn) A fact or event that is observed.

precipitation (prih-sih-puh-TAY-shuhn) Water that falls to the ground as hail, mist, rain, sleet, or snow.

technology (tehk-NAH-luh-jee) A method that uses science to solve problems and the tools used to solve those problems.

tropical (TRAH-pih-kuhl) Of or relating to the tropics, usually having to do with warm, wet weather.

tundra (TUHN-druh) A large area of flat land in northern parts of the world where there are no trees and the ground is always frozen.

violently (VY-uh-luhnt-lee) With very strong force.

water cycle (WAH-tuhr SY-kuhl) The process through which water in various forms passes through the atmosphere to Earth's surface and back into the atmosphere.

INDEX

PRIMARY SOURCE LIST

Page 12
"Johnstown Flood, May 31st, 1889. No. 15, General view of Johnstown looking towards Kernville." Photograph. Ernest Walter. 1889. Now kept at the Library of Congress Prints and Photographs Division, Washington, D.C.

Page 15
Environmental engineer inspects the dry cracked floor of Pejar Dam. Photograph. Greg Wood. July 8, 2005. New South Wales, Australia.

Page 17
A heavily damaged area in the aftermath of a tornado in Rowlett, Texas. Photograph. Laura Buckman. December 27, 2015. Rowlett, Texas.

WEBSITES

Due to the changing nature of Internet links, PowerKids Press has developed an online list of websites related to the subject of this book. This site is updated regularly. Please use this link to access the list: www.powerkidslinks.com/swnd/world